Standard Measure
States, Great Britain

Arthur S. C. Wurtele

Alpha Editions

This edition published in 2024

ISBN : 9789362097743

Design and Setting By
Alpha Editions
www.alphaedis.com
Email - info@alphaedis.com

As per information held with us this book is in Public Domain.
This book is a reproduction of an important historical work. Alpha Editions uses the best technology to reproduce historical work in the same manner it was first published to preserve its original nature. Any marks or number seen are left intentionally to preserve its true form.

Contents

INTRODUCTION. ....- 1 -

STANDARD MEASURES. ....- 2 -

UNITED STATES. ....- 5 -

FRANCE. ....- 7 -

COMPARISONS OF UNITED STATES
AND ENGLISH STANDARDS. ....- 12 -

COMPARISON OF UNITED STATES
AND FRENCH STANDARDS. ....- 14 -

COMPARISON OF ENGLISH AND
FRENCH STANDARDS. ....- 17 -

APPENDIX. ....- 22 -

INTRODUCTION.

During the preparation of this investigation of Standard Measures a large number of authorities were examined, including the following: Kelly's "Universal Cambist," Maunder's "Weights and Measures," "Encyclopædia Britannica," "Chambers' Encyclopædia," Williams' "Geodesy," Hymer's works, "Smithsonian Reports," "Coast Survey Reports," Herschel's "Astronomy," etc. The only concise and clear statement I found was J. E. Hilgard's report to the Coast Survey on standards in 1876, which I was gratified to find coincides with my deductions.

ARTHUR S. C. WURTELE.

ALBANY, November 26, 1881.

STANDARD MEASURES.

A standard measure of length at first sight appears to be very simple—merely a bar of metal of any length, according to the unit of any country; and comparisons of different standards do not seem to present any difficulty. But on looking further into the thing, we find that standards are referred to some natural invariable length, and we are at once confronted with a mass of scientific reductions giving different values to the same thing, according to successively improved means of observation. We find, also, that comparisons of one standard with another differ, as given by reductions carried to great apparent exactness.

Every author appears to assume the right of using his own judgment as to what reduction is to be considered the most exact, and the result is a very confusing difference in apparently exact figures, with nothing to show how these differences arise.

I have endeavored to indicate what may be the cause of this confusion by giving the figures of actually observed comparisons and reductions; in a manner, the roots of the figures used as statements of length.

Sir Joseph Whitworth gives 1/40000 of an inch as the smallest length that can be measured with certainty, with an ultimate possibility of 1/1000000 of an inch; but imperceptible variations of temperature affect these infinitesimal lengths to such an extent that he believes the limit can only be reached at a standard temperature of 85° F., to avoid the effect of heat of the body.

It appears to me that comparisons should be made of double yards and mètres with the old French toise, as the limit of exactness would be thereby doubled.

Another great defect in statements of relative values is the omission of necessary facts—the material of which the bars or standards are made, the temperature at which comparison was made, and the standard temperatures used as to the final reduction, with the coefficient of expansion adopted.

Again, bars of different metals appear in time to sensibly change their relative length.

ENGLISH STANDARDS OF LENGTH.

The first establishment of a uniform standard appears to have been made in 1101 by Henry I., who is said to have fixed the ulna (now the yard) at the length of his arm; but nothing definite was done till 1736, when the

Royal Society took steps toward securing a general standard, and in 1742 they had a standard yard made by Graham from a comparison of various yards and ells of Henry VII. and Elizabeth, that were kept in the Exchequer.

Two copies of the Royal Society standard yard were made by Bird in 1758 for a committee of Parliament, one of which was marked "standard of 1758," and the other 1760. But no exact legal standard was yet established, as shown by comparisons in 1802 of the various standard measures in use which Pictet, of Geneva, made with an accurate scale by Troughton, using means exact to the ten thousandth part of an inch, with the following results at the temperature of 62° F.:

Troughton Scale		36·00000	inches.
Parliamentary Standard (1758, Bird)		36·00023	"
Royal Society "	(1760, ")	35·99955	"
" "	(Graham)	36·00130	"
Exchequer "		35·99330	"
Tower "		36·00400	"
Gen. Roy "	(Trig. Survey)	36·00036	"

Parliament finally undertook to reform the measures of England, and appointed a commission in 1818, under whose authority Capt. Kater compared the standard yards then in use with the following results, as referred to the Indian Survey standard:

Col. Lambton Standard (Indian Survey)	36·000000	inches.
Bird's Standard (1760)	36·000659	"
Sir Geo. Schuckburgh's Standard	36·000642	"
Ramsden's Bar. Ordnance Survey	36·003147	"
Gen. Roy's Scale	36·001537	"
Royal Society Standard	36·002007	"

The commission reported in favor of adopting Bird's standard of 1760, as it differed so slightly from Sir George Schuckburgh's standard (which had been used in deducing the value of the French mètre) that those values could be assumed as correct. They also established the length of the seconds pendulum at level of sea in London and in vacuo as 39·13929 inches. The seconds pendulum had been previously fixed by Wollaston and Playfair in 1814 as 39·13047 inches.

On this report, an Act of Parliament in 1823 declared the only standard measure of length for the United Kingdom to be the yard as given by the distance at 32° F. between two points in gold studs on the brass bar, made by Bird, and marked "Standard of 1760," and in the keeping of the Clerk of the House of Commons; also it referred this standard yard to the natural standard of a pendulum vibrating seconds of mean solar time at the level of the sea, in vacuo at London and temperature of 32° F., as in the proportion of 36 to 39·13929; so that a pendulum 36 inches long ought to make 90088·42 vibrations in 24 hours.

The Royal Society had a copy of the legal standard made by Bailey in 1834; and in the same year the Parliamentary standard was destroyed by fire at the burning of the Houses of Parliament, leaving the kingdom again without a legal standard.

All attempts made by a commission consisting of Airy, Bailey, Herschel, Lubbock, and Sheepshanks, to restore the standard by means of the seconds pendulum failed in exactness, on account of the many conditions of a vibrating pendulum, and recourse was had to the Royal Society standard, which had been carefully compared by Captain Kater in 1818, and from this in 1838 Bailey and Sheepshanks made six bronze bars, one inch square, and 38 inches long, which in 1855 were legalized by Act of Parliament, and the English standard of length defined as follows:

"That the straight line on distance between the centres of the transverse lines in the two gold plugs on the bronze bar deposited in the Exchequer shall be the genuine standard yard at the temperature of 62° Fahrenheit; and if lost, it shall be replaced by means of its copies."

The French metrical system was made legal permissively in 1864, at the length established by Captain Kater, referred to in Act of Parliament of 1823, of 1 mètre equal to 39·37079 inches, or 3·28089916 feet.

These are the standards now in use in the United Kingdom.

UNITED STATES.

By the Constitution of the United States Congress is charged with fixing the standard of measures (Art. 1, sec. 8); but as no enactment has been made by Congress, the standard yard in England, which was legal previous to 1776 in the Colonies, is the standard yard of the United States, and does not differ with the English standard yard.

Under resolution of Congress in 1830, Mr. Hassler was employed to examine the standards in use.

Considerable discrepancies were found, but the mean of all examined corresponded very nearly with the English standard, and in 1832 the recommendation of Mr. Hassler was adopted, and the standard yard defined as the distance between the 27th and 63d inch marks, at the temperature of 62° F., on the brass scale 82 inches long, being an exact copy of Sir George Schuckburgh's standard, made by Troughton, of London, for the Coast Survey, and deposited in the Office of Weights and Measures at Washington.

In 1836 an Act of Congress ordered standards to be sent to each Governor of a State, and the work was done under direction of Mr. Hassler.

In 1856, two copies of the English standard yard, as restored after destruction of the original standard by fire in 1834, No. 11 of bronze, and No. 57 of Low Moor wrought iron, were presented to the United States by Airy.

The United States Troughton standard bar being compared with No. 11 was found to be longer by 0·00085 inch, or in proportion of 1 to 1·0000237216, about 1½ inches in a mile, according to Report of Secretary of Treasury in 1857.

Later comparisons made by J. E. Hilgard, of the Coast Survey, at the British Standards Office, between No. 11 and the standard imperial yard, give No. 11 as 0·000088 inch shorter, or it would be of standard length at temperature of 62·25° F.

We may infer that the Troughton standard is too long by 0·000762 inch, or would be standard length at temperature 59·77° F. instead of at 62° by making expansion reduction with Airy's coefficient for the bronze of the imperial standards, 0·000342 inch per yard for 1° F.

The mètre was made a legal standard permissively in 1866; the United States mètre standard being one of the 12 iron mètre bars made and verified for the French Government in 1799 on the adoption of the metrical system, and brought to America by Mr. Hassler in 1800, the relative value being fixed by Act of Congress at 39·37 inches.

The relative value of 39·36850154 United States inches, as obtained by Mr. Hassler, corrected to 62° F., was used by the Coast Survey till 1868, when it was found advisable to use the relative value of 39·3704 as deduced by Clarke. Since 1800 several standard mètre bars were sent to the United States by the French Government, and on comparison, there appearing to be a slight discrepancy, the original iron standard mètre bar was sent to Dr. F. A. P. Barnard in Paris, and in 1867 it was compared with the French platinum standard, which is only used once in ten years to verify other standards.

A difference was found by this comparison of only ·00017 millimètre or 1/160000 inch, which being only 1/100 of an inch in a mile is inappreciable.

FRANCE.

The standard of length of the système ancient was the toise of 6 pieds, divided into 12 pouces of 12 lignes each.

The origin of the toise is not known, but it was probably legally established by Philip Le Bel, about 1300, as he first appears to have taken steps toward a uniform system of measures in France. In the 13th century the toise is mentioned by Ch. Le Rains. In the 14th century Menongier writes that, in marching, the sight should strike the ground 4 toises in front. In the fifteenth century Pereforest brings in the toise, and in the sixteenth century the Contume de Berry says, "We use in this country two toises; one for carpenters of 5 pieds and a half, the other for masons of 6 pieds."

Picard used the toise in his measurement of an arc of meridian from Malvoisin to London in 1669.

The meridians measured by the Academy in 1735 to settle the question of the figure of the earth were made by means of two standard toises, known as the "Toise du Nord," and the "Toise du Sud."

The first, used by Maupertuis, Clairault, and Le Monnier, in Lapland, was destroyed by immersion in sea-water, when their ship was wrecked on the return voyage.

The second, with which La Condamine, Bourgner, and Godin operated in Peru, was the original of the toise Canivet made in 1768, and of the standards used in determining the mètre.

The commencement of the move for a scientific standard of length in France which resulted in the mètre was in 1790, when the revolutionary government proposed to England the formation of a commission of equal numbers from the English Royal Society and the French Academy, for the purpose of fixing the length of the seconds pendulum at latitude 45° as the basis of a new system of measures. This proposal was not favorably received, and the Academy, at the request of government, appointed as a commission Borda, Lagrange, Laplace, Monge, and Condorcet, to decide whether the seconds pendulum, the quarter of the equator, or the quarter of a meridian, should be used as the natural standard for the new system of measures. They settled on the last as best for the purpose, and resolved that the ten millionth of the meridian quadrant, or distance from equator to pole, measured at sea level, be taken for basis of the new system, and be called a mètre.

Delambre and Mechin were at once charged with re-measurement of the meridian surveyed in 1739 by La Caille and Cassini, from Dunkirk to Perpignan, and its extension to Barcelona.

Operations were commenced in 1792, and carried on with great accuracy to completion in 1799; Delambre working between Dunkirk and Paris, and Mechin between Paris and Barcelona.

The distance measured from Dunkirk to Barcelona was 9° 40′ 24·24″ of arc, or 1,075,059 mètres, as reduced to the new standard.

The "toise de Peru" was the standard used in the work at a temperature of 13° R.

Two base-lines were measured with Borda's compensating bars of brass and platinum; one at Melun, near Paris, 6076 toises long, and the second at Perpignan, 6028 toises long, and though over 900,000 mètres apart, the calculated length differed by only 10 pouces.

This meridian was afterward, in 1806, extended by Gen. Roy to Greenwich, on the north, and by Biot and Arago to Formentera, on the south. The results, as given by Laplace in centesimal degrees and mètres, are as follows:

Greenwich	57·19753°	·0 mètres.
Pantheon, Paris	54·27431°	292,719·3 "
Formentera	42·96178°	1,423,636·1 "

The middle of the arc being 50·079655° Cent., or 45° 4′ 18·0822″ Sexa., and the middle degree centesimal being very nearly 100,000 mètres.

The determination of the final result of these geodetic measurements was referred to a committee of 20 members; 9 named by the French Government, and the others by the governments of Holland, Savoy, Denmark, Spain, Tuscany, and of the Cisalpine, Ligurian, and Swiss republics, on the invitation of France.

This committee established the meridian quadrant at 5,130,740 toises; making the mètre 0·513074 of the toise, or 36·9413 pouces, or 443·296 lignes, and the toise 1·94903659 mètres.

Iron standard mètre bars, 12 in number were made by Borda, also 2 of platinum and 4 standard toise bars.

The 12 standard iron mètre bars were sent to different countries, after being verified by the French Government, and on the 2d of November,

1801, the mètrical système was legalized by France, and the standard unit of length declared to be the ten millionth part of a meridian quadrant of the earth, as defined by the distance at a temperature of 0° Centigrade (32° F.) between two points on a platinum bar in the keeping of the Academy of Science at Paris. This standard bar is used only once every ten years for exact comparisons, as stated by Dr. F. A. P. Barnard.

About 1837 Bessel, by a combination of 11 measured arcs of meridian, deduced the quadrant of meridian as 5,131,179·81 toises instead of 5,130,740 toises, as fixed by law. This would make to quadrant 10,000,565·278 legal mètres, or would increase the mètre length from 443·296 lignes to 443·334 lignes, agreeing very nearly with result obtained by Airy in 1830, from a combination of 13 measured arcs.

The following are the measured arcs used by Bessel and Airy; the combinations being indicated by initial letters, A and B.

	Measurer.	*Mid. Lat.*	*Arc.*	*Length.*
B.—	Svanberg, Sweden	+ 66° 20′ 10·0″	1° 37′ 19·6′	593,277 feet
A.—	Maupertuis, Sweden	+ 66° 19′ 37·0″	0° 57′ 30·4′	351,832 "
A.—	Struve, Russia	+ 58° 17′ 37·0″	3° 35′ 5·2″	1,309,742 "
B.—	Struve and Tenner, Russia	+ 56° 3′ 55·5″	8° 2′ 28·9′	2,937,439 "
B.—	Bessel and Bayer, Prussia	+ 54° 58′ 26·0″	1° 30′ 29·0′	551,073 "
B.—	Schumacher, Denmark	+ 54° 8′ 13·7″	1° 31′ 53·3′	559,121 "
A, B. —	Ganss, Hanover	+ 52° 32′ 16·6″	2° 0′ 57·4′	736,425 "
A.—	Roy and Kater, England	+ 52° 35′ 45·0″	3° 57′ 13·1′	1,442,953 "

B.—	" " "		+ 52°	2′	19·0″	2°	50′	23·5′	1,036,409 "
A.—	Lacaille and Cassini, France		+ 46°	52′	2·0″	8°	20′	0·3″	3,040,605 "
A, B.	Delambre and Mechin, France		+ 44°	51′	2·5″	12°	22′	12·7′	4,509,832 "
A.—	Boscovich, Rome		+ 42°	59′	·0″	2°	9′	47·0′	787,919 "
A.—	Mason and Dixon, America		+ 39°	12′	·0″	1°	28′	45·0′	538,100 "
A, B.	Lambton, India		+ 16°	8′	21·5″	15°	57′	40·7′	5,794,598 "
A, B.	Lambton and Everest, India		+ 12°	32′	20·8″	1°	34′	56·4′	574,318 "
A, B.	Lacondamine, Peru		− 1°	31′	0·4″	3°	7′	3·5″	1,131,050 "
A.—	Lacaille, Cape Good Hope		− 33°	18′	30·0″	1°	13′	17·5′	445,506 "
B.—	Maclear, " "		− 35°	43′	20·0″	3°	34′	34·7′	1,301,993 "
A.—	Plana and Cartessi, Piedmont		——— —			1°	7′	31·1′	——— —

The following different lengths of the mètre have been obtained:

As adopted by France, 1801   443·296   lignes.

According to Delambre   443·264   "

" Bessel   443·33394 "

" Airy	443·32387	"
" Clarke	443·36146	"
From Peru Meridian	443·440	"

The length of a pendulum vibrating 100,000 times in a mean solar day was determined in numerous careful experiments by Biot, Arago, and Mathieu, in mètres of 443·296 lignes, as follows:

Dunkirk	56·67	lat. Cent.	0	above sea	0·7419076	mètres.
Paris	54·26	"	65	"	0·7418870	"
" by Borda	54·26	"	0	"	0·7416274	"
Bordeau	49·82	"	0	"	0·7412615	"
Formentera	42·96	"	196	"	0·7412061	"

Borda also determined the length of the seconds pendulum at Paris, in vacuo:

First result	440·5595	lignes	= 0·9938267	mètre.
Second result	"	"	= 0·9938460	"
As given by Ganot	"	"	= 0·9935	"

In 1812 the système usuelle was established, of which the unit was one third of the mètre, with the old name of pied, and duodecimally divided into pouces and lignes.

This system continued in use till 1840, when it was abolished by law, and the names of pied, pouce, and ligne forbidden under penalties. So the mètre, decimally divided, remains the only legal measure of length in France.

COMPARISONS OF UNITED STATES AND ENGLISH STANDARDS.

In 1832, under resolution of Congress, Mr. Hassler compared the different standard yards in America, with the following results, using the yard between the twenty-seventh and sixty-third inches on the scale made of bronze by Troughton, of London, for the United States Coast Survey, as the reference, that being identical with Sir George Schuckburg's standard:

Troughton Scale, mid. yard	36·0000000 inches.
" " between platinum points	35·9989758 "
Jones yard in State Department	35·9990285 "
Iron yard in Engineer Department	35·9987760 "
Brass yard, Albany, Sec. of State	36·0002465 "
Gilbert yard, University of Virginia	35·9952318 "

In 1856 the Troughton standard bronze scale was compared with the bronze standard yard No. 11, which was sent over by Airy as a copy of the English imperial standard, as restored after destruction of the original standard by fire in 1834, and the United States standard was found to be longer by 0·00085 inch.

Later comparisons by J. E. Hilgard, of the Coast Survey, of the bronze standard No. 11 with the imperial standard yard, at the British Standards Office, gave No. 11 as 0·000088 shorter than the imperial standard.

Hassler's reduction of the mètre, as deduced by Beach at 62° F., 39·36850154, compared with the English reduction of the mètre, 39·37079 inches, gives an excess to the United States Standard of 0·002029 inch.

The following reductions have been given for the United States yard in English inches:

Report of Sec. of Treas., 1857	36·00087	= 1·00002416
Chambers' Encyclopædia, 1872	36·00087	
" " "	36·0020892	= 1·0000580334
Trautwine	36·0020894	= 1·000058038
Mathewson, U. S. surveyor	36·00208944	= 1·00005804

Hassler and Beach	36·002092	= 1·00005811
J. E. Hilgard, Coast Survey	36·00076	= 1·000021

To Mr. Hassler's reduction the name of United States inch has been applied; but his reduction is not correct, as he used a rate of expansion for brass deduced by himself of 0·0003783 inch in one yard for 1° F., and later experiments show that the smaller rate of 0·000342, deduced by Airy, is more correct.

By correcting Hassler's reduction with the later rate of expansion, J. E. Hilgard shows that the difference would be very small, or only 36·0002286 = 1·00000635, or about ⅖ of an inch in a mile.

In Coast Survey report for 1876, J. E. Hilgard calls attention to another difficulty in the matter of extreme accuracy, in the uncertainty with regard to the permanence in the length of a bar, and states that the bronze standard bar No. 11 and the Low Moor iron standard bar No. 57, presented to the United States by Great Britain, are found to have changed their relative length by 0·00025 inch in 25 years; the bronze bar being now relatively shorter by that amount. This subject, he states, is undergoing further investigation.

COMPARISON OF UNITED STATES AND FRENCH STANDARDS.

In 1817 Mr. Hassler examined the French standards in America, for the Coast Survey, using the Troughton bronze standard scale, which is identical with Sir George Schuckburg's standard, as the reference, with the following results, all being reduced to temperature of 32° F.

Original Iron Mètre,	1799	39·381022708	inches.
Lenoir Iron Mètre,	Coast Survey	39·37972015	"
" Brass "	"	39·380247972	"
" " "	Eng. Dept.	39·38052739	"
Canivet Iron Toise,	1768	76·74334472	"
Lenoir " "		76·74192710	"

In 1814 Troughton had compared with his own scale in London two of the above.

Lenoir Iron Mètre, C. S. 39·3802506 inches.

" Brass "   "   39·3803333   "

In 1832, under resolution of Congress, Hassler again compared the French standards in the United States, using as before the Troughton scale, and reducing all to temperature of 32° F. as follows:

Original	Iron Mètre,	1799	39·3808643	inches.
Lenoir	"	" C. S.	39·3799120	"
"	Brass Mètre	C. S.	39·380447	"
"	"	Eng. Dept.	39·3801714	"
"	"	" in 1829	39·3807095	"
Fortin	"	State Dept.	39·3796084	"
	"	Treas. "	39·3795983	"
	Iron Mètre	" "	39·3807827	"

Gilbert	"	Univ. of Virg.		39·365408	"
	Platinum Mètre			39·3803278	"
	"	(Nicollet)		39·380511	"
Canivet	Iron Toise,		1768	76·74290511	"
Lenoir	"		1799	76·74047599	"

From the mean of his comparisons between the United States brass Troughton standard yard and the authentic French standard mètres used by the Coast Survey, Hassler, in 1832, deduced the value of the mètre at 39·3809172 inches, at 32° F., and by correction for expansion to United States standard temperature of 62° F., he made the mètre at 32° equal to 39·36850154 inches at 62° F.

The British imperial standard and the United States Troughton standard differ by only 0·000762 inch, which applied to the English reduction of 39·37079, would give 39·36996 as the relative value according to Troughton standard.

The difference between these reductions is probably to be attributed to the use of different rates of expansion, in correcting for standard temperatures, which vary considerably, according to high authority as follows for brass at 1° F.

Whitworth, 1876	0·00000956	= 0·00034416 in. per yard.
Borda, 1799	0·000009913	= 0·00035687 "
Smeaton, 1750	0·000010417	= 0·00037501 "
Hassler	0·000010508	= 0·0003783 "
Ramsden, 1760	0·000010516	= 0·0003786 "
Faraday, 1830	0·00001059	= 0·00038124 "

And for the bronze of which the British imperial standards are made:

| Airy and Sheepshanks | 0·0000095 | = 0·000342 in. per yard. |
| Fizeau | 0·00000975 | = 0·000351 " |

The correction at Ramsden's rate is nearly identical with Hassler's, and gives 39·3684933; at Whitworth's rate it would give 39·36962, very nearly the same as deduced from the difference between the British Imperial standard and the United States Troughton standard. The results of Sir

Joseph Whitworth were obtained by use of all late improvements for scientific precision, and they must be accepted as most reliable.

It would appear preferable to give comparisons at the same temperature in connection with the corrected result, so that international comparisons of scientific measurements may not be vitiated by accidental variations.

COMPARISON OF ENGLISH AND FRENCH STANDARDS.

When the mètre standard was established in France, 1799, it was compared with Sir George Schuckburg's standard yard by Captain Kater. The quadrant of 10,000,000 mètres, or 5,130,740 toises, was determined to be 32,808,992 English feet, giving the mètre equal to 3·2808992 English feet, or 39·37079 inches, and the toise equal to 6·3945925921 English feet.

In 1814 Wollaston and Playfair, by comparison with the platinum mètre standard at 55° F., deduced the mètre as equal to 39·3828 English inches.

During the geodetic operations of General Roy in 1802, who used 60° F. as standard temperature, Pictet's comparisons, using means capable of measuring the 10,000th part of an inch, gave the mètre standard, which is used at 32° F. as standard temperature, at 39·3828 English inches; this corrected for temperature by Dr. Young, gave 39·371 English inches at 62° F.; which result was confirmed by Bird, Maskelyne and Laudale.

In 1823, by Act of Parliament on report of committee, the mètre is fixed as 39·37079 English inches.

In 1800 the Royal Society, by comparison with two toise standards sent by Lalande to Maskelyne, deduced the mètre as 39·3702 English inches.

Later comparisons by Clarke in the Ordnance Survey Office at Southampton, in 1866, give the mètre as 39·37043 inches.

The French Academy of Sciences by comparison with Sir George Schuckburg's standard at temperature of 32° F., deduced the mètre as 39·3824 English inches, which reduced to standard temperature of 62° F., would be 39·3711, or slightly in excess of the value deduced by Dr. Young from Pictet's comparisons.

The legal value in England is one mètre equal to 39·37079, and the latest reduction is 39·37043 inches by Clarke in 1866, which is probably the most exact reduction.

DIFFERENT REDUCTIONS OF THE FRENCH TOISE INTO ENGLISH FEET.

Captain Kater, 1799	6·3945925921	feet.
Hassler, 1832	6·3951409	"

Chambers' Encyclopædia	6·39456	"
" Mathematics	6·394662	"
Wallace	6·39462	"
Nystrom	6·39625	"
Alexander	6·39435	"
Dana	6·3946	"

The following table of reductions as used shows clearly how great a confusion exists in the matter of comparisons:

MÈTRE IN INCHES.

Phœnixville Hand-book	39·368	inches.
Hassler	39·36850154	"
"	39·370788	"
"	39·3809172	"
Trautwine	39·368505	"
"	39·37079	"
Silliman	39·368505	"
"	39·37079	"
Chambers' Encyclopædia	39·36850535	"
" "	39·3707904	"
Act of United States Congress, 1866	39·37	"
Smithsonian Report	39·37	"
Youmans	39·37	"
Davies	39·37	"
Homan's Encyclopædia	39·37008	"
Weale	39·3702	"
Ordnance Survey (England, 1866)	39·37043	"
Clerk Maxwell	39·37043	"

Capt. Clarke	39·3704316	"
J. M. Rankine (1870)	39·3704316	"
" (1866)	39·3707904	"
Alexander (weights and measures)	39·37068	"
Ganot	39·370788	"
Vose	39·370788	"
Act of British Parliament, 1823	39·37079	"
Encyclopædia Britannica	39·37079	"
Hymer	39·37079	"
Davies and Peck	39·37079	"
J. W. Clarke	39·37079	"
Dana	39·37079	"
Whittaker	39·37079	"
Sommerville	39·3707904	"
Chambers' Mathematics	39·3707904	"
Gwilt's Encyclopædia	39·3707904	"
Gillespie	39·3707904	"
Capt. Kater	39·3708	"
Appleton's Encyclopædia	39·37079	"
Van Nostrand	39·3708	"
D'Aubuisson	39·3708	"
Johnson (draftsman)	39·3708	"
Encyclopædia Americana	39·371	"
Jameson's Dictionary	39·371	"
Herbert's Encyclopædia	39·371	"
Popular "	39·371	"
Molesworth	39·371	"

Dr. Young (1802)	39·371	"
Wallace (engineer)	39·371	"
Nystrom	39·38091	"
Hencke	39·3809172	"
Act of Canadian Parliament, 1873	39·3819	"
Paris Academy	39·3824	"

LENGTH OF THE SECONDS PENDULUM AS GIVEN BY DIFFERENT WRITERS.

NEW YORK.—	Hencke	39·1012	inches.
	Bartlet	39·11256	"
	Nystrom	39·1017	"
	Ganot	39·1012	"
	Byrne	39·10153	"
	Wallace	39·10153	"
LONDON.—	Hencke	39·13908	"
	Gillespie	39·13929	"
	Chambers' Encyclopædia	39·13929	"
	Williams' Geodesy	39·13929	"
	Act of Parliament, 1823	39·13929	"
	Wallace (engineer)	39·1393	"
	Chambers' Mathematics	39·1393	"
	Hymer Astronomy	39·13734	"
	Bartlet	39·13908	"
	Vose	39·1393	"
	Sommerville	39·1393	"
	Nystrom	39·1393	"
	Davies and Peck	39·13908	"

	Ganot	39·1398	"
	Wollaston (1814)	39·13047	"
	Galbraith	39·139	"
	Byrne	39·1393	"
	Capt. Kater	39·13829	"
PARIS.—	Hencke	39·12843	"
	Ganot	39·1285	"
	Galbraith	39·128	"
	Byrne	39·12843	"
	Wallace	39·12843	"

APPENDIX.

Having shown in the preceding pages that in the point of view of scientific accuracy the yard, mètre, and toise standards are on a common level, and that in the matter of comparisons there is no extreme accuracy, I will now refer to the proposed change of our standard from the yard to the mètre.

Theoretically the mètre is the 10,000,000th part of the earth's quadrant, and the yard the 36/39·13929th part of a seconds pendulum at London. Practically, neither the mètre nor yard could be recovered with exactness from their natural basis. The legal French mètre differs from the latest reduction enough to give an excess of over three miles to the circumference of the earth. In fact, the mètre and yard are only the lengths of bars of metal kept in certain offices, from which copies are made. Decimally considered, it is as easy to divide one as the other into tenths, hundredths, etc., and the yard standard is often so divided.

As to nomenclature, the metrical system is overloaded with Greek and Latin prefixes, which are in no way so easy and convenient in expression as the short, sharp Anglo-Saxon words yard, foot, inch.

In all sciences Latin and Greek names are given for easier purposes of classification; but the different peoples invariably keep their own household names for daily purposes, leaving prefix and affix to specialists, probably with advantage to both parties.

The units used for different purposes are entirely distinct from the base of any system, and though always referable to such base, are not practically so referred. It therefore seems useless to burden the people with long scientific names in the ordinary transactions of daily life.

For long distances the units in the yard and metrical systems are respectively the mile and the kilomètre.

The mile has a definite meaning in our minds, being associated, from the days of youth, with the measured distances in race-courses, speed in walking, railway and steamer travel, length of surveyed lots—the same being in use among about 100,000,000 people.

For mechanical structures, the units are respectively the foot and the mètre. The foot is used instead of the yard, as being the most convenient in practice, and is fixed in the minds of the people by constant association with length of foot-rules, size of buildings, doors, windows, etc., all of which are always before us.

For commercial purposes the units are respectively the yard and the metre. The yard is associated with length of yard-sticks, distance between brass nails on counters, so many finger-lengths by ladies. Probably three fourths of the business of the world is conducted on the yard standard.

For machine and shop work the English unit is the inch and fractions, and countries having the metrical standard have universally adopted the millimètre.

The inch is well fixed in the minds of all mechanics by constant use, and the ease with which the fractions are had by halving only renders the system very convenient.

As more figures must be used to indicate a size by millimètres than by inches and fractions, it appears that the metrical system cannot shorten the work of arithmetical computation in shop work, and is therefore of no advantage to the mechanic or draftsman, but rather the reverse. This is the opinion of Coleman Sellers, the distinguished Philadelphia engineer and manufacturer, who, after a trial of the millimètre in his shops for some years, returned to the use of the inch, and writes in *Engineering News*: "The loss from the use of a small unit requiring many figures to express what is needed, takes away from the other advantages of the system when considered from a labor-saving point of view."

In France itself the metrical system is not wholly decimal in actual practice, as we find the following measures in use in addition to the decimal divisions: double decamètre, demi-decamètre, double mètre, demi-mètre, and double decimètre.

The metrical system has been adopted in the following countries: France and colonies, Holland and colonies, Belgium, Spain and colonies, Portugal, Italy, Germany, Greece, Roumania, British India, Mexico, New Granada, Ecuador, Peru, Brazil, Uruguay, Argentine Confederacy, Chili, Venezuela; and partially in Wurtemburg, Bavaria, Baden, Hesse, Switzerland, Denmark, Austria, and Turkey.

In the past centuries all the work and records of English-speaking peoples—now numbering about 100,000,000, and increasing and progressing faster than all other nationalities, as well as being closely connected by descent and business—have been done and recorded under the yard standard, and any change now would inevitably render necessary continual reductions, to the great detriment and inconvenience of the mass of our people, and with little or no practical benefit, except perhaps to a small class of scientific and pseudo-scientific men, who can and do amuse themselves with the fancied uniformity of the mètre.

All our numerous text-books and tables, mechanical and scientific, would be rendered entirely useless by the change, and this is a serious final consideration.